D0563694

WHAT WE KNOW ABOUT CLIMATE CHANGE

WHAT WE KNOW ABOUT CLIMATE CHANGE

Kerry Emanuel

Afterword by Judith A. Layzer and
William R. Moomaw

A Boston Review Book

THE MIT PRESS Cambridge, Mass. London, England

MIT Press books may be purchased at special quantity
discounts for business or sales promotional use. For
information, please e-mail special_sales@mitpress.mit.edu or
write to Special Sales Department, The MIT Press,
55 Hayward Street, Cambridge, MA 02142.

This book was set in Adobe Garamond by *Boston Review*
and was printed and bound in the United States of America.

Designed by Joshua J. Friedman

Library of Congress Cataloging-in-Publication Data
Emanuel, Kerry A., 1955–
    What we know about climate change / Kerry Emanuel.
        p.   cm. — (Boston Review books)
    ISBN: 978-0-262-05089-0 (hardcover : alk. paper)
    I. Global warming. I. Title.
QC981.8.G56E43 2007
363.738'74—dc22   2007010953

CONTENTS

WHAT WE KNOW ABOUT CLIMATE CHANGE

1

Two strands of environmental philosophy run through the course of human history. The first holds that the natural state of the universe is one of infinite stability, with an unchanging earth anchoring the predictable revolutions of sun, moon, and stars. Every scientific revolution that challenged this notion, from Copernicus' heliocentricity to Hubble's expanding universe, from Wegener's continental drift to Heisenberg's uncertainty and Lorenz's macroscopic chaos, met with fierce resistance from religious, political, and even scientific hegemonies.

The second strand also sees the natural state of the universe as a stable one but holds that it has become destabilized through human actions. The great floods are usually portrayed in religious traditions as attempts by a god or gods to cleanse the earth of human corruption. Deviations from cosmic predictability, such as meteors and comets, were more often viewed as omens than as natural phenomena. In Greek mythology, the scorching heat of Africa and the burnt skin of its inhabitants were attributed to Phaeton, an offspring of the sun god Helios, who, having lost a wager to his son, was obliged to allow him to drive the sun chariot across the sky. In this primal environmental catastrophe, Phaeton lost control and fried the earth, killing himself in the process.

These two fundamental ideas have permeated many cultures through much of history. They strongly influence views of climate change to the present day.

In 1837, Louis Agassiz provoked public outcry and scholarly ridicule when he proposed that many puzzles of the geologic record, such as peculiar scratch marks on rocks and boulders far removed from their bedrock sources, could be explained by the advance and retreat of huge sheets of ice. This event marked the beginning of a remarkable endeavor, today known as paleoclimatology, which uses physical and chemical evidence from the geological record to deduce changes in the earth's climate over time. This undertaking has produced among the most profound yet least celebrated scientific advances of our era. We now have exquisitely detailed knowledge of how climate has varied over the last few million years and, with progressively less detail and more uncertainty, how it has changed going back in time to the age of the oldest rocks on our 4.5-billion-year-old planet.

For those who take comfort in stability, there is little consolation in this record. Within the past three million years or so, our climate has swung between mild states, similar to today's and lasting from 10,000 to 20,000 years, and periods of 100,000 years or so in which giant ice sheets, in some places several miles thick, covered northern continents. Even more unsettling than the existence of these cycles is the suddenness with which the climate can apparently change, especially as it recovers from glacial eras.

Over longer intervals of time, the climate has changed even more radically. During the early part of the Eocene era, around 50 million years ago, the earth was free of ice, and giant trees grew on islands near the North Pole, where the annual mean temperature was about 60°F, far warmer than today's mean of about 30. There is also some evidence that the earth was almost entirely covered with ice at various times around 500 million years ago; in between, the planet was exceptionally hot.

What explains these changes? For climate scientists, the ice cores in Greenland and Antarctica provide the most intriguing clues. As the ice formed, it trapped bubbles of atmosphere, whose chemical composition—including, for example, its carbon dioxide and methane content—can now be analyzed. Moreover, it turns out that the ratio of the masses of two isotopes of oxygen locked up in the molecules of ice is a good indicator of the air temperature when the ice was formed. And to figure out when the ice was formed, one can count the layers that mark the seasonal cycle of snowfall and melting.

Relying on such analyses of ice cores and sediment cores from the deep ocean, climate scientists have learned something remarkable: the ice-age cycles of the past three million years are probably caused by periodic oscillations of the earth's orbit that primarily affect the orientation of the earth's axis. These oscillations do not much affect the *amount* of sunlight that reaches the earth, but they do change the *distribution* of sunlight with latitude. This distribution matters because land and water absorb and reflect sunlight differently, and the distributions of land and water—continents and oceans—are quite different in the northern and southern hemispheres. Ice ages occur when, as a result of orbital variations, the arctic regions intercept relatively little summer sunlight so that ice and snow do not melt as much.

The timing of the ice ages, then, is the combined result of the earth's orbit and its basic geology. But this combination does not explain either the slow pace of the earth's descent into the cold phases of the cycle or the abrupt recovery to interglacial warmth evident in the ice-core records. More disturbing is the evidence that these large climate swings—from glacial to interglacial and back—are caused by relatively small changes in the distribution of sunlight with latitude. Thus, on the time scale of ice ages, climate seems exquisitely sensitive to small perturbations in the distribution of sunlight.

And yet for all this sensitivity, the earth never suffered either of the climate catastrophes of fire or ice. In the fire scenario, the most effective greenhouse gas—water vapor—accumulates in the atmosphere as the earth warms. The warmer the atmosphere, the more water vapor can accumulate; as more water vapor accumulates, more heat gets trapped, and the warming spirals upward. This uncontrolled feedback is called the runaway greenhouse effect, and it continues until the oceans have all evaporated, by which time the planet is unbearably hot.

One has only to look as far as Venus to see the end result. Any oceans that may have existed on that planet evaporated eons ago, yielding a super greenhouse inferno and an average surface temperature of around 900°F.

Death by ice can result from another runaway feedback. As snow and ice accumulate progressively equatorward, they reflect an increasing amount of sunlight back to space, further cooling the planet until it freezes into a "snowball earth." It used to be supposed that once the planet reached such a frozen state, with almost all sunlight reflected back to space, it could never recover; more recently it has been theorized that without liquid oceans to absorb the carbon dioxide continuously emitted by volcanoes, that gas would accumulate in the atmosphere until its greenhouse effect was finally strong enough to start melting the ice.

It would not take much change in the amount of sunlight reaching the earth to cause one of these catastrophes. And solar physics informs us that the sun was about

25 percent dimmer early in the earth's history, which should have led to an ice-covered planet, a circumstance not supported by geological evidence.

So what saved the earth from fire and ice?

Life itself may be part of the answer to the riddle of the faint young sun. Our atmosphere is thought to have originated in gases emitted from volcanoes, but the composition of volcanic gases bears little resemblance to air as we know it today. It is thought that the early atmosphere consisted mostly of water vapor, carbon dioxide, sulfur dioxide, chlorine, and nitrogen. There is little evidence that there was much oxygen—until the advent of life. The first life forms helped produce oxygen through photosynthesis and transformed the atmosphere into something like today's, consisting mostly of nitrogen and oxygen with trace amounts of water vapor, carbon dioxide, methane, and other gases. Carbon-dioxide content probably decreased slowly with time owing to chemical weathering, possibly aided by biological processes. As the composition changed, the net greenhouse effect weakened, compensating for the slow but inexorable brightening of the sun.

Thus early life dramatically changed the planet. We humans are only the most recent species to do so.

The compensation between increasing solar power and decreasing greenhouse effect may not have been an accident. In the 1960s, James Lovelock proposed that life actually exerts a stabilizing influence on climate by producing feedbacks favorable to itself. He called his idea the Gaia hypothesis, named after the Greek earth goddess.

But even according to this view, life is only preserved in the broadest sense: individual species, such as those that transformed the early atmosphere, altered the environment at their peril.

2

As this sketch of the planet's early climatic history shows, the greenhouse effect plays a critical role in the earth's climate, and no sensible discussion of climate could proceed without grasping its nature. (A cautionary note: the greenhouse metaphor itself is flawed. Whereas actual greenhouses work by preventing convection currents from carrying away heat absorbed from sunlight, the atmosphere prevents heat from *radiating* away from the surface.)

The greenhouse effect has to do with radiation, which in this context refers to energy carried by electromagnetic waves, which include such phenomena as visible light, radio waves, and infrared radiation. All matter with a temperature above absolute zero emits radiation. The hotter the substance, the more radiation it emits and the shorter the average wavelength of the radiation emitted. A fairly narrow range of wavelengths constitute visible light. The average surface temperature of the sun is about 10,000°F, and the sun emits much of its radiation as visible light, with an average wavelength of about half a micron. (A micron is one millionth of a meter; there are 25,400 microns in an inch.) The earth's atmosphere emits as though its average temperature were around 0°F, at an average wavelength of about 15 microns. Our eyes cannot detect this infrared radiation. It is important to recognize that the same object can both emit and absorb radiation: when an object emits radiation it loses energy, and this has the effect of cooling it; absorption, on the other hand, heats an object.

Most solids and liquids absorb much of the radiation they intercept, and they also emit radiation rather easily. Air is another matter. It is composed almost entirely of oxygen and nitrogen, each in the form of two identical atoms bonded together in a single molecule. Such molecules barely interact with radiation: they allow free passage to both solar radiation moving downward to the earth and infrared radiation moving upward from the earth's surface. If that is all there were to the atmosphere, it would be a simple matter to calculate the average temperature of the earth's surface: it would have to be just warm enough to emit enough infrared radiation to balance the shortwave radiation it absorbed from the sun. (Were it too cool, it would emit less radiation than it absorbed and would heat up; conversely, were it too warm it would cool.) Accounting for the amount of sunlight reflected back to space by the planet, this works out to be about 0°F, far cooler than the observed mean surface temperature of about 60°F.

Fortunately for us, our atmosphere contains trace amounts of other substances that do interact strongly with radiation. Foremost among these is water, H_2O, consisting of two atoms of hydrogen bonded to a single atom of oxygen. Because of its more complex geometry, it absorbs and emits radiation far more efficiently than molecular nitrogen and oxygen. In the atmosphere, water exists both in its gas phase (water vapor) and its condensed phase (liquid water and ice) as clouds and precipitation. Water vapor and clouds absorb sunlight and infra-red radiation, and clouds also reflect sunlight back to space. The amount of water vapor in a sample of air varies greatly from place to place and time to time, but in no event exceeds about two percent of the mass of the sample. Besides water, there are other gases that interact strongly with radiation; these include CO_2, or carbon dioxide (presently about 380 tons for each million tons of air), and CH_4, or methane (around 1.7 tons for each million tons of air).

Collectively, the greenhouse gases are nearly transparent to sunlight, allowing the short-wavelength radiation to pass virtually unimpeded to the surface, where much of it is absorbed. (But clouds both absorb and reflect sunlight.) On the other hand, these same gases absorb much of the long-wavelength, infrared radiation that passes through them. To compensate for the heating this absorption causes, the greenhouse gases must also emit radiation, and each layer of the atmosphere thus emits infrared radiation upward and downward.

As a result, the surface of the earth receives radiation from the atmosphere as well as the sun. It is a remarkable fact that, averaged over the planet, the surface receives more radiation from the atmosphere than directly from the sun! To balance this extra input of radiation—the radiation emitted by atmospheric greenhouse gases and clouds—the earth's surface must warm up and thereby emit more radiation itself. This is the essence of the greenhouse effect.

If air were not in motion, the observed concentration of greenhouse gases and clouds would succeed in raising the average temperature of the earth's surface to around

85°F, much warmer than observed. In reality, hot air from near the surface rises upward and is continually replaced by cold air moving down from aloft; these convection currents lower the surface temperature to an average of 60°F while warming the upper reaches of the atmosphere. So the emission of radiation by greenhouse gases keeps the earth's surface warmer than it would otherwise be; at the same time, the movement of air dampens the warming effect and keeps the surface temperature bearable.

3

THIS BASIC CLIMATE PHYSICS IS EN-
tirely uncontroversial among scientists. And
if one could change the concentration of a
single greenhouse gas while holding the rest
of the system (except its temperature) fixed,
it would be simple to calculate the corresponding change in surface temperature.
For example, doubling the concentration of
CO_2 would raise the average surface temperature by about 2.1°F, enough to detect
but probably not enough to cause serious
problems. Almost all the controversy arises
from the fact that in reality, changing any

single greenhouse gas will indirectly cause other components of the system to change as well, thus yielding additional changes. These knock-on effects are known as feedbacks, and the most important and uncertain of these involves water.

A fundamental difference exists between water and most other greenhouse gases. Whereas a molecule of carbon dioxide or methane might remain in the atmosphere for hundreds of years, water is constantly recycled between the atmosphere, land surface, and oceans, so that a particular molecule of water resides in the atmosphere for, on average, about two weeks. On climate time scales, which are much longer than two weeks, atmospheric water is very nearly in equilibrium with the surface, which means that as much water enters the atmosphere by evaporating from the surface as is lost to the surface by rain and snow. One cannot simply tally up the sources and sinks and figure out which wins; a more involved argument is needed.

To make matters worse, water vapor and clouds are far and away the most important greenhouse substances in the atmosphere, and clouds also affect climate not only by sending infrared radiation back to earth and warming it but by reflecting sunlight back into space, thus cooling the planet. Water is carried upward from its source at the surface by convection currents, which themselves are a byproduct of the greenhouse effect, which tends to warm the air near the surface. Simple physics as well as detailed calculations using computer models of clouds show that the amount of water vapor in the atmosphere is sensitive to the details of the physics by which tiny cloud droplets and ice crystals combine into larger raindrops and snowflakes, and how these in turn fall and partially re-evaporate on their way to the surface. The devil in these details seems to carry much authority with climate.

This complexity is limited, however, because the amount of water in the atmosphere is subject to a fundamental and important constraint. The concentration of water vapor in any sample of air has a strict upper limit that depends on its temperature and pressure: in particular, this limit rises very rapidly with temperature. The ratio of the actual amount of water vapor in a sample to this limiting amount is the familiar quantity called *relative humidity*. Calculations with a large variety of computer models and observations of the atmosphere all show that as climate changes, relative humidity remains approximately constant. This means that as atmospheric temperature increases, the actual amount of water vapor increases as well. But water vapor is a greenhouse gas. So increasing temperature increases water vapor, which leads to further increases in temperature. This positive feedback in the climate system is the main reason why the global mean surface temperature is expected to increase somewhat more than the 2.1°F that doubling CO_2 would produce in the absence of feedbacks. (At very high temperatures, the water vapor feedback can run away, leading to the catastrophe of a very hot planet, as mentioned before.)

The amount and distribution of water vapor in the atmosphere is also important in determining the distribution of clouds, which play a complex role in climate. On the one hand, they reflect about 22 percent of the incoming solar radiation back to space, thereby cooling the planet. On the other hand, they absorb solar radiation and both absorb and emit infrared radiation, thus contributing to greenhouse warming. Different global climate models produce wildly different estimates of how clouds might change with changing climate, thus constituting the largest source of uncertainty in climate-change projections.

A further complication in this already complex picture comes from anthropogenic aerosols—small solid or liquid particles suspended in the atmosphere. Industrial activity and biomass burning have contributed to large increases in the aerosol content of the atmosphere, and this is thought also to have had a large effect on climate.

The main culprits are the sulfate aerosols, which are created through atmospheric chemical reactions involving sulfur dioxide, another gas produced by the combustion of fossil fuels. These tiny particles reflect incoming sunlight and, to a lesser degree, absorb infrared radiation. Perhaps more importantly, they also serve as condensation nuclei for clouds. When a cloud forms, water vapor does not form water droplets or ice crystals spontaneously but instead condenses onto pre-existing aerosol particles. The number and size of these particles determines whether the water condenses into a few large droplets or many small ones, and this in turn strongly affects the amount of sunlight that clouds reflect and the amount of radiation they absorb.

It is thought that the increased reflection of sunlight to space—both directly by the aerosols themselves and through their effect on increasing the reflectivity of clouds—outweighs any increase in their greenhouse effect, thus cooling the planet. Unlike the greenhouse gases, however, sulfate aerosols only remain in the atmosphere a few weeks before they are washed out by rain and snow. Their abundance is proportional to their rate of production—as soon as production decreases, sulfate aerosols follow suit. Since the early 1980s, improved technology and ever more stringent regulations have diminished sulfate aerosol pollution in the developed countries, aided by the collapse of the USSR and the subsequent reduction of industrial output there. On the other hand, sources of sulfate aerosols have been steadily increasing in Asia and the developing countries, so it is unclear how the net global aerosol content has been changing over the past 25 years.

Important uncertainties enter the picture, then, with water (especially clouds) and airborne particulates. But the uncertainties actually go much deeper: indeed, to understand long-term climate change, it is essential to appreciate that detailed forecasts cannot, *even in principle*, be made beyond a few weeks. That is because the climate system, at least on short time scales, is *chaotic*.

The essential property of chaotic systems is that small differences tend to magnify rapidly. Think of two autumn leaves that have fallen next to each other in a turbulent brook. Imagine following them as they move downstream on their way to the sea: at first, they stay close to each other, but the eddies in the stream gradually separate them. At some point, one of the leaves may get temporarily trapped in a whirlpool behind a rock while the other continues downstream. It is not hard to imagine that one of the leaves arrives at the mouth of the river days or weeks ahead of the other. It is also not hard to imagine that a mad scientist, having equipped our brook with all kinds of fancy instruments for measuring the flow of water and devised a computer program for predicting where the leaves would go, would find it almost impossible to predict where the leaf would be even an hour after it started its journey.

Let's go back to the two leaves just after they have fallen in the brook, and say that at this point they are ten inches apart. Suppose that after 30 minutes they are ten feet apart, and this distance increases with time. Now suppose that it were possible to rewind to the beginning but this time start the leaves only five inches apart. It would not be surprising if it took longer—say an hour—before they are once again 10 feet apart. Keep rewinding the experiment, each time decreasing the initial distance between the leaves. You might suppose that the time it takes to get 10 feet apart keeps increasing indefinitely. But for many physical systems (probably including brooks), this turns out not to be the case. As you keep decreasing the initial separation, the increases in the amount of time it takes for the leaves to be separated by 10 feet get successively smaller, so much so that there is a definite limit: no matter how close the leaves are when they hit the water, it will not take longer than, say, six hours for them to be ten feet apart.

The same principle applies if, instead of having two leaves, we have a single leaf and a computer model of the leaf and the stream that carries it. Even if the computer model is perfect and we start off with a perfect representation of the state of the brook, any error—even an infinitesimal one—in the timing or position of the leaf when it begins its journey will lead to the forecast being off by at least ten feet after six hours, and greater distances at longer times. Prediction beyond a certain time is impossible.

Not all chaotic systems have this property of limited predictability, but our atmosphere and oceans, alas, almost certainly do. As a result, it is thought that the upper limit of the predictability of weather is around two weeks. (That we are not very close to this limit is a measure of the imperfection of our models and our measurements.)

While the day-to-day variations of the weather are perhaps the most familiar examples of environmental chaos, variations at longer time scales can also behave chaotically. El Niño is thought to be chaotic in nature, making it difficult to predict more than a few months in advance. Other chaotic phenomena involving the oceans have even longer time scales, but beyond a few years it becomes increasingly difficult for scientists to tell the difference between chaotic natural variations and what climate scientists called "forced" variability. But this difference is important for understanding the human role in producing climate change.

On top of the natural, chaotic "free" variability of weather and climate are changes brought about by changing "forcing," which is usually considered to involve factors that are not themselves affected by climate. The most familiar of these is the march of the seasons, brought about by the tilt of the earth's axis, which itself is independent of climate. The effects of this particular forcing are not hard to separate from the background climate chaos: we can confidently predict that January will be colder than July in, say, New York. Other examples of natural climate forcing include variations in solar output, and volcanic eruptions, which inject aerosols into the stratosphere and thereby cool the climate.

Some of this natural climate forcing is chaotic in nature, but some of it is predictable on long time scales. For example, barring some catastrophic collision with a comet or asteroid, variations of the earth's orbit are predictable many millions of years into the future. On the other hand, volcanic activity is unpredictable. In any event, the actual climate we experience reflects a combination of free (unforced), chaotic variability, and changes brought about by external forcing, some of which, like volcanic eruptions, are themselves chaotic. And part of this forced climate variability is brought about by us human beings.

4

An important and difficult issue in detecting anthropogenic climate change is telling the difference between natural climate variations—both free and forced—and those that are forced by our own activities.

One way to tell the difference is to make use of the fact that the increase in greenhouse gases and sulfate aerosols dates back only to the industrial revolution of the 19th century: before that, the human influence is probably small. If we can estimate how climate changed before this time, we will have some idea of how the system varies naturally.

Unfortunately, detailed measurements of climate did not themselves really begin in earnest until the 19th century; but there are "proxies" for quantities like temperature, recorded in, for example, tree rings, ocean and lake plankton, pollen, and corals.

Plotting the global mean temperature derived from actual measurements and from proxies going back a thousand years or more reveals that the recent upturn in global temperature is truly unprecedented: the graph of temperature with time shows a characteristic hockey-stick shape, with the business end of the stick representing the upswing of the last 50 years or so. But the proxies are imperfect and associated with large margins of error, so any hockey-stick trends of the past may be masked, though the recent upturn stands above even a liberal estimate of such errors.

Another way to tell the difference is to simulate the climate of the last 100 years or so with climate models. Computer modeling of global climate is perhaps the most complex endeavor ever undertaken by mankind. A typical climate model consists of millions of lines of computer instructions designed to simulate an enormous range of physical phenomena, including the flow of the atmosphere and oceans; condensation and precipitation of water inside clouds; the transfer of solar and terrestrial radiation through the atmosphere, including its partial absorption and reflection by the surface, by clouds, and by the atmosphere itself; the convective transport of heat, water, and atmospheric constituents by turbulent convection currents; and vast numbers of other processes. There are by now a few dozen such models in the world, but they are not entirely independent of one another, often sharing common pieces of computer code and common ancestors.

Although the equations representing the physical and chemical processes in the climate system are well known, they cannot be solved exactly. It is computationally impossible to keep track of every molecule of air and ocean, and to make the task viable, the two fluids must be divided up into manageable chunks. The smaller and more numerous these chunks, the more accurate the result, but with today's computers the smallest we can make these chunks in the atmosphere is around 100 miles in the horizontal and a few hundred yards in the vertical, and a bit smaller in the ocean. The problem here is that many important processes are much smaller than these scales. For example, cumulus clouds in the atmosphere are critical for transferring heat and water upward and downward, but they are typically only a few miles across and so cannot be simulated by the climate models. Instead, their effects must be represented in terms of the quantities like wind and temperature that pertain to the whole computational chunk in question.

The representation of these important but unresolved processes is an art form known by the awful term *parameterization*, and it involves numbers, or parameters, that must be tuned to get the parameterizations to work in an optimal way. Because of the need for such artifices, a typical climate model has many tunable parameters, and this is one of many reasons that such models are only approximations to reality. Changing the values of the parameters or the way the various processes are parameterized can change not only the climate simulated by the model, but the sensitivity of the model's climate to, say, greenhouse-gas increases.

How, then, can we go about tuning the parameters of a climate model in such a way as to make it a reasonable facsimile of reality? Here important lessons can be learned from our experience with those close cousins of climate models, weather-prediction models. These are almost as complicated and must also parameterize key physical processes, but because the atmosphere is measured in many places and quite frequently, we can test the model against reality several times per day and keep adjusting its parameters (that is, tuning it) until it performs as well as it can. But with climate, there are precious few tests. One obvious hurdle the model must pass is to be able to replicate the current climate, including key aspects of its variability, such as weather systems and El Niño. It must also be able to simulate the seasons in a reasonable way: the summers must not be too hot or the winters too cold, for example.

Beyond a few simple checks such as these, there are not too many ways to test the model, and projections of future climates must necessarily involve a degree of faith. The amount of uncertainty in such projections can be estimated to some extent by comparing forecasts made by many different models, with their different parameterizations (and, very likely, different sets of coding errors). We operate under the faith that the real climate will fall among the projections made with the various models; in other words, that the truth will lie somewhere between the higher and lower estimates generated by the models.

The figure on the facing page shows the results of two sets of computer simulations of the global average surface temperature of the 20th century using a particular climate model. In the first set, denoted by the darker shade of gray, only natural, time-varying forcings are applied; these consist of variable solar output and "dimming" owing to aerosols produced by known volcanic eruptions. The second set (lighter gray) adds in the man-made influences on sulfate aerosols and greenhouse gases. In each set, the model is run four times beginning with slightly different initial states, and the range among the four ensemble members is denoted by the shading in the figure, reflecting the free random variability of the climate produced by this model, while the colored curves show the average of the four ensemble members. The observed global

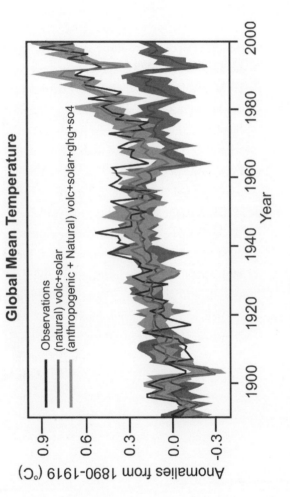

Global Mean Temperature

Observations
(natural) volc+solar
(anthropogenic + Natural) volc+solar+ghg+so4

Anomalies from 1890-1919 (°C)

Year

average surface temperature is depicted by the black curve. One observes that the two sets of simulations diverge during the 1970s and have no overlap at all today, and that the observed global temperature also starts to fall outside the envelope of the all-natural simulations in the 1970s. This exercise has been repeated using many different climate models with the same qualitative result: one cannot simulate the evolution of the climate over last 30 years without including in the simulations mankind's influence on sulfate aerosols and greenhouse gases. This, in a nutshell, is why almost all climate scientists today believe that man's influence on climate has emerged from the background noise of natural variability.

5

PROJECTIONS BASED ON CLIMATE MODELS suggest that the globe will continue to warm another 3 to 7°F over the next century. This is similar to the temperature change one could experience by moving, say, from Boston to Philadelphia. Moreover, the warming of already hot regions—the tropics—is expected to be somewhat less, while the warming of cold regions like the arctic is projected to be more, a signal already discernible in global temperature measurements. Nighttime temperatures are increasing more rapidly than daytime warmth.

Is this really so bad? In all the negative publicity about global warming, it is easy to overlook the benefits: It will take less energy to heat buildings, previously infertile lands of high latitudes will start producing crops, and there will be less suffering from debilitating cold waves. Increased CO_2 might also make crops grow faster. On the down side, there will be more frequent and more intense heat waves, air conditioning costs will rise, and previously fertile areas in the subtropics may become unarable. Sure, there will be winners and losers, but will the world really suffer in the net? Even if the changes we are bringing about are larger than the globe has experienced in the last few thousand years, they still do not amount to the big natural swings between ice ages and interglacial periods, and the earth and indeed human beings survived these.

But there are consequences of warming that we cannot take so lightly. During the peak of the last ice age, sea level was some 400 feet lower than it is today, since huge quantities of water were locked up in the great continental ice sheets. As polar regions warm, it is possible that portions of the Greenland and Antarctic ice sheets will melt, increasing sea level. Highly detailed and accurate satellite-based measurements of the thickness of the Greenland ice show that it is actually increasing in the interior but thinning around the margins, and while there are also patterns of increase and decrease in Antarctic ice, it appears to be thinning on the whole. Meltwater from the surface of the Greenland ice sheet is making its way to the bottom, possibly allowing the ice to flow faster toward the sea. Our understanding of the physics of ice under pressure is poor, and it is thus difficult to predict how the ice will respond to warming. Were the entire Greenland ice cap to melt, sea level would increase by around 22 feet—flooding many coastal regions including much of southern Florida and lower Manhattan.

My own work has shown that hurricanes are responding to warming sea surface temperatures faster than we originally expected, especially in the North Atlantic, where the total power output by tropical cyclones has increased by around 60 percent since the 1970s. The 2005 hurricane season was the most active in the 150 years of records, corresponding to record warmth of the tropical Atlantic. Hurricanes are far and away the worst natural disasters to affect the U.S. in economic terms. Katrina may cost us as much as $200 billion, and it has claimed at least 1,200 lives. Globally, tropical cyclones cause staggering loss of life and misery. Hurricane Mitch of 1998 killed over 10,000 people in Central America, and in 1970 a single storm took the lives of some 300,000 people in Bangladesh. Substantial changes in hurricane activity cannot be written off as mere climate perturbations to which we will easily adjust.

Basic theory and models show another consequential result of a few degrees of warming. The amount of water vapor in the air rises exponentially with temperature: a seven-degree increase in temperature increases water vapor by 25 percent. One might at first suppose that since the amount of water ascending into clouds increases, the amount of rain that falls out of them must increase in proportion. But condensing water vapor heats the atmosphere, and in the grand scheme of things, this must be compensated by radiative heat loss. On the other hand, simple calculations show that the amount of radiative heat loss increases only very slowly with temperature, so that the total heating by condensation must increase slowly as well. Models resolve this conundrum by making it rain harder in places that are already wet and at the same time increasing the intensity, duration, or geographical extent of droughts. Thus, the twin perils of flood and drought actually both increase substantially in a warmer world.

It is particularly sobering to contemplate such outcomes in light of the evidence that smaller, natural climate swings since the end of the last ice age debilitated and in some cases destroyed entire civilizations in such places as Mesopotamia, Central and South America, and the southwestern region of what is today the United States.

In pushing the climate so hard and so fast, we are also conscious of our own collective ignorance of how the climate system works. Perhaps negative-feedback mechanisms that we have not contemplated or have underestimated will kick in, sparing us from debilitating consequences. On the other hand, the same could be said of positive feedbacks, and matters might turn out worse than projected. The ice-core record reveals a climate that reacts in complex and surprising ways to smoothly and slowly changing radiative forcing caused by variations in the earth's orbit. Far from changing smoothly, it remains close to one state for a long time and then suddenly jumps to another state. We do not understand this, and are worried that a sudden climate jump may be part of our future.

6

SCIENCE PROCEEDS BY CONTINUALLY testing and discarding or refining hypotheses, a process greatly aided by the naturally skeptical disposition of scientists. We are, most of us, driven by a passion to understand nature, but that means being dispassionate about pet ideas. Partisanship—whatever its source—is likely to be detected by our colleagues and to yield a loss of credibility, the true stock of the trade. We share a faith—justified by experience—that at the end of the day, there is a truth to be found, and those who cling for emotional reasons to wrong ideas will be judged by history accordingly, whereas those who see it early will be regarded as visionaries.

The evolution of the scientific debate about anthropogenic climate change illustrates both the value of skepticism and the pitfalls of partisanship. Although the notion that fossil-fuel combustion might increase CO_2 and alter climate originated in the 19th century, general awareness of the issue dates to a National Academy of Sciences report in 1979 that warned that doubling CO_2 content might lead to a three-to-eight-degree increase in global average temperature. Then, in 1988, James Hansen, the director of NASA's Goddard Institute for Space Studies, set off a firestorm of controversy by testifying before Congress that he was virtually certain that a global-warming signal had emerged from the background climate variability. At that time, less was known about natural climate variability before the beginning of systematic instrumental records in the 19th century, and only a handful of global climate simulations had been performed. Most scientists were deeply skeptical of Hansen's claims; I certainly was. It is important to interpret the word "skeptical" literally here: it was not that we were sure of the opposite, merely that we thought the jury was out.

At roughly this time, radical environmental groups and a handful of scientists influenced by them leapt into the fray with rather obvious ulterior motives. This jumpstarted the politicization of the issue, and conservative groups, financed by auto makers and big oil, responded with counterattacks. This also marked the onset of an interesting and disturbing phenomenon that continues to this day: a very small number of climate scientists adopted dogmatic positions, and in so doing lost credibility among the vast majority who remained committed to an unbiased search for answers.

On the left, an argument emerged urging fellow scientists to deliberately exaggerate their findings to galvanize an apathetic public, an idea that, fortunately, failed in the scientific arena but took root in Hollywood, culminating in the 2004 release of *The Day After Tomorrow*. On the right, the search began for negative feedbacks that would counter increasing greenhouse gases; imaginative ideas emerged, but they have largely failed the acid test of comparison to observations.

But as the dogmatists grew increasingly alienated from the scientific mainstream, they were embraced by political groups and journalists, who thrust them into the limelight. This produced a gross distortion in the public perception of the scientific debate. Ever eager for the drama of competing dogmas, the media largely ignored mainstream scientists, whose hesitations did not make good copy. As the global-warming signal continues to emerge, this soap opera is kept alive by a dwindling number of deniers constantly tapped for interviews by journalists who pretend to look for balance.

While the American public has been misinformed by a media obsessed with sensational debate, climate scientists have developed a way forward that helps them to compare notes and test one another's ideas, and also creates a valuable communication channel. Called the Intergovernmental Panel on Climate Change, or IPCC, it produces a detailed summary of the state of the science every four years or so; the most recent report was published in April 2007. Although far from perfect, the IPCC involves serious climate scientists from many countries and has largely withstood political attack and influence.

The IPCC reports are fairly candid about what we collectively know and where the uncertainties probably lie. In the first category are findings that are not in dispute, not even by *les refusards*:

⁋ Concentrations of the greenhouse gases carbon dioxide, methane, ozone, and nitrous oxide are increasing owing to fossil-fuel consumption and biomass burning. Carbon dioxide has increased from its pre-industrial level of about 280 parts per million (ppmv) to about 380 ppmv today, an increase of about 35 percent. From ice-core records, it is evident that present levels of CO_2 exceed those experienced by the planet at any time over at least the past 650,000 years.

¶ Concentrations of certain anthropogenic aerosols have also increased owing to industrial activity.

¶ The earth's average surface temperature has increased by about 1.2°F in the past century, with most of the increase occurring from about 1920 to 1950, and again beginning around 1975. The years 1998 and 2005 were the warmest in the instrumental record.

¶ Sea level has risen by about 2.7 inches over the past 40 years; of this, a little over an inch occurred during the past decade.

¶ The annual mean geographical extent of arctic sea ice has decreased by 15 to 20 percent since satellite measurements of this began in 1978.

61

In the second category are findings that most climate scientists agree with but that are disputed by some:

¶ The global mean temperature is now greater than at any other time in at least the past 500 to 1,000 years.

¶ Most of the global mean temperature variability is caused by four factors: variability of solar output, major volcanic eruptions, and anthropogenic sulfate aerosols and greenhouse gases.

¶ The dramatic rise in global mean temperature in the past 30 years is owing primarily to increasing greenhouse-gas concentrations and a leveling off or slight decline in sulfate aerosols.

¶ Unless measures are taken to reduce greenhouse-gas production, global mean temperature will continue to increase, about 2.5 to 9°F over the next century, depending on uncertainties and how much greenhouse gas is produced.

¶ As a result of the thermal expansion of sea water and the melting of polar ice caps, sea level will increase six to 16 inches over the next century, though the increase could be larger if large continental ice sheets become unstable.

¶ Rainfall will continue to become concentrated in increasingly heavy but less frequent events.

¶ The incidence, intensity, and duration of both floods and drought will increase.

¶ The intensity of hurricanes will continue to increase, though their frequency may dwindle.

All these projections depend, of course, on how much greenhouse gas is added to the atmosphere over the next century, and even if we could be certain about the changes, estimating their net effect on humanity is an enormously complex undertaking, pitting uncertain estimates of costs and benefits against the costs of curtailing greenhouse-gas emissions. But we are by no means certain about what kind of changes are in store, and we must be wary of climate surprises. Even if we believed that the projected climate changes would be mostly beneficial, we might be inclined to make sacrifices as an insurance policy against potentially harmful surprises.

7

ESPECIALLY IN THE UNITED STATES, the political debate about global climate change became polarized along the conservative–liberal axis some decades ago. Although we take this for granted now, it is not entirely obvious why the chips fell the way they did. One can easily imagine conservatives embracing the notion of climate change in support of actions they might like to see anyway. Conservatives have usually been strong supporters of nuclear power, and few can be happy about our current dependence on foreign oil. The United States is renowned for its technological innovation and should be at an advantage in making money from any global sea change in energy-producing technology: consider the prospect of selling new means of powering vehicles and electrical generation to China's rapidly expanding economy. But none of this has happened.

Paradoxes abound on the political left as well. A meaningful reduction in greenhouse-gas emissions will require a shift in the means of producing energy, as well as conservation measures. But such alternatives as nuclear and wind power are viewed with deep ambivalence by the left. Senator Edward Kennedy, by most measures our most liberal senator, is strongly opposed to a project to develop wind energy near his home in Hyannis, and environmentalists have only just begun to rethink their visceral opposition to nuclear power. Had it not been for green opposition, the United States today might derive most of its electricity from nuclear power, as does France; thus the environmentalists must accept a large measure of responsibility for today's most critical environmental problem.

There are other obstacles to taking a sensible approach to the climate problem. We have preciously few representatives in Congress with a background or interest in science, and some of them display an active contempt for the subject. As long as we continue to elect scientific illiterates like Senator James Inhofe, who believes global warming to be a hoax, we will lack the ability to engage in intelligent debate. Scientists are most effective when they provide sound, impartial advice, but their reputation for impartiality is severely compromised by the shocking lack of political diversity among American academics, who suffer from the kind of groupthink that develops in cloistered cultures. Until this profound and well-documented intellectual homogeneity changes, scientists will be suspected of constituting a leftist think tank.

On the bright side, the governments of many countries, including the United States, continue to fund active programs of climate research, and many of the critical uncertainties about climate change are slowly being whittled down. The extremists are being exposed and relegated to the sidelines, and when the media stop amplifying their views, their political counterparts will have nothing left to stand on. When this happens, we can get down to the serious business of tackling the most complex and perhaps the most consequential problem ever confronted by mankind.

Like it or not, we have been handed Phaeton's reins, and we will have to learn how to control climate if we are to avoid his fate.

Judith A. Layzer
and William R. Moomaw

AFTERWORD

DESPITE THE COMPLEXITY AND UN-predictability of the global climate system, there are factors that make some futures far more likely than others. In particular, we know that society's introduction of more heat-trapping gases into the atmosphere will almost certainly lead to a warmer world, higher sea levels, and more intense droughts and storms. Furthermore, because half of each ton of carbon dioxide we emit today will be in the atmosphere a century from now, and because the thermal momentum of the oceans and the melting of glaciers we have already set in motion will continue for 3,000 years even if we stop burning fossil fuels immediately, some damages are inevitable. We also know that concerted efforts to reduce emissions and enhance the ability of terrestrial ecosystems to absorb carbon dioxide can minimize the rise in global temperature, thereby dampening the most severe consequences of global warming. Given the high probability of extremely adverse outcomes and our ability to forestall them, a prudent person would conclude that we should act now. Why, then, is the United States moving so slowly—and how might we change course?

The main reason for our political inertia is that proponents of policies to address global warming have struggled to translate climate-change science into a politically compelling story, while their opponents have effectively shifted attention to the potential costs of addressing the problem. For many years the U.S. environmental community lacked the elements of a narrative that could capture the public imagination: the villains were ordinary Americans, and the most affected victims were small island nations; the relationship between heat-trapping gases and global temperature was complex—mediated by many variables and amplified or dampened by highly uncertain feedbacks; and the crisis, should there be one, appeared to be at least a century away. Opponents countered this already weak narrative with a persuasive alternative storyline. They emphasized the uncertainties in climate-change science, actively supporting a handful of contrarians. As important, they claimed that instituting policies to curb emissions of carbon dioxide and other greenhouse gases would cripple the American economy. For a decade, these arguments—which were widely disseminated thanks to enormous infusions of cash from fossil-fuel-based industries—succeeded in defusing public concern.

Over the last five years, however, scientists have provided a steady stream of research that strengthens the global-warming story and decisively discredits the contrarians. First, a more visible and increasingly certain international scientific consensus about humans' impact on the global climate has rendered absurd claims that scientists are divided. As the IPCC points out, in recent years the cause-effect relationship between human-caused carbon-dioxide emissions and rising global temperature has emerged unmistakably from the statistical noise. Scientists have corrected divergent satellite temperature measurements, quantified most climate-forcing factors, and tested and rebutted the most plausible alternative explanations for the observed temperature rise.

Second, scientists have sought to detect and forecast regional impacts of a changing climate and thereby highlight the extent to which Americans not only are the perpetrators but also will be the victims of global warming. They have generated scenarios that reveal the enormous local costs of regional climate changes; for example, the Northeast will experience severe flooding as the Atlantic Ocean rises; California will suffer severe disruptions in its water supplies as snow packs diminish; and throughout the West, droughts will become longer and more frequent, and wildfires will become more numerous and severe. As for the crisis, research indicates that it draws closer by the day: scientists are already documenting

75

changes in the nesting and mating habits of species around the world and faster-than-expected melting of polar ice caps and glaciers from Greenland to Antarctica and tropical glaciers from the Andes to Kilimanjaro. Moreover, scientists are detecting unanticipated impacts of additional carbon dioxide, such as increases in the ocean's acidity and phytoplankton declines that promise to be disastrous for marine ecosystems.

Less publicized but as important is the likelihood that addressing global warming could be relatively painless. It is true that to maintain concentrations of atmospheric carbon dioxide low enough to keep the global temperature from rising three times the amount it already has (1.2°F)—an increase many scientists believe will destabilize the climate in dangerous ways—we must reduce global emissions by 75 percent or more.

But although that figure sounds overwhelming, we can achieve it if the United States and other industrial economies reduce their emissions by three percent per year between now and mid-century. Continuing at this rate until the end of the century will bring our emissions down by nearly 95 percent, and as a result atmospheric carbon-dioxide concentrations will begin to fall back to today's levels.

Technological solutions are necessary but insufficient to reduce emissions by three percent per year; we will need to make lifestyle changes as well. But most of those adjustments will be negligible—and many will yield multiple benefits. For the American who drives 1,000 miles each month—the national average—driving 30 miles less per month for a year constitutes a three-percent reduction for that year. Most drivers could save those miles by occasionally sharing a ride to work or taking public transit. Others could achieve equivalent savings by driving less aggressively. Even better, by replacing an SUV with a fuel-efficient vehicle, a driver can instantaneously cut her emissions in half—the equivalent of an annual three-percent savings for 23 years. Similarly, it is relatively simple to reduce emissions from most existing buildings by 30 percent simply by adding insulation and energy-efficient lighting; we can make even greater reductions by replacing old appliances and installing modern windows and furnaces or ground-source heat pumps.

These changes are unlikely to come about in response to market forces alone; fortunately, however, as decades of experience with environmental regulation demonstrates, putting in place a set of policies that establish consistent and predictable rules can spur both rapid technological innovation and behavior change. As a first step, we should dismantle the web of policies that overwhelmingly favors fossil-fuel production and use and actively discriminates against new technologies and practices that would reduce harmful emissions. We routinely subsidize fossil fuels by allowing mining companies to extract coal by blowing off the tops of mountains and dumping the waste into Appalachian rivers; streamlining permits to develop oil and gas on publicly owned territory in the Rocky Mountain West and offshore Alaska; and using military force to prop up oil-producing regimes around the world. Similarly, policies that protect large, obsolete coal-burning power plants in the United States obstruct efforts to make a transition to newer, more

efficient power sources, including renew-
ables and distributed, combined heat and
power systems.

The second step is to institute federal,
state, and local policies that reverse the dis-
incentives created by the existing policy
structure and force users to pay the costs of
extracting, transporting, and burning fossil
fuels. The most straightforward and effective policy changes would include a carbon
tax; an increase in the corporate average
fuel economy (CAFE) standards; and a large
increase in funding for mass transit, both
within cities and along heavy travel routes
on the East and West coasts. A less obvious
policy change would be to require those who
introduce energy-consuming technologies
to offset or save one and a half times the
amount of new emissions generated. State
and local governments can adopt growth-

management policies that reflect the environmental costs of sprawling and inefficient development—such as upgrading building codes to ever-tightening Energy Star standards for renovations and new construction; creating incentives to increase urban densities and redevelop inner-city brownfields; downzoning rural areas; and putting areas of critical environmental concern, such as coastal and freshwater wetlands, off limits to development.

In deciding which technologies and behaviors to encourage, we will need to depart from our past practice of treating each remedy in isolation and instead think at a systems level. For example, widespread use of biofuels may reduce emissions from power plants and vehicles, but if their production entails clearing additional land or using more fertilizer, we could negate any benefits by eliminating carbon sinks and producing more heat-trapping nitrous oxide. Similarly, although some commentators have touted nuclear energy as a straightforward solution to global warming, no one has yet developed a credible plan for storing highly radioactive waste or dealing with the very real threats of natural disasters, technological failure, or the use of nuclear technology by terrorists or hostile states. In short, when choosing from the menu of available policy tools, we should give top priority to those that encourage reducing consumption and adopting technologies that minimize rather than shift environmental impacts.

Of course, devising effective policies is much easier than implementing them. Enacting major policy change entails political risk and is likely only if aspiring leaders perceive substantial public concern—and therefore the possibility of political support for their stands. Fortunately, although the public has been slow to react to the threat of global warming, public opinion—like the climate system—is subject to tipping points, and there is abundant evidence that the United States is nearing one. Actions taken at the state and local levels not only attest to widespread public concern but are also triggering positive economic and political feedbacks that strengthen demands for national policies.

Meanwhile, the British Parliament is seriously considering a climate-change bill that would mandate a 60 percent cut in the nation's carbon emissions by 2050, and the European Union recently announced a unilateral reduction in CO_2 emissions of 20 percent by 2020. China and India—whose emissions are growing rapidly, although per capita they remain dramatically lower than those in the developed world—are also taking aggressive steps in the right direction. In 2008, China's vehicle standards will exceed those of the United States by a substantial margin in each vehicle class. Although China's coal-burning power plants are only 33 percent efficient—compared to an advanced design potential of around 45 percent—they nevertheless perform better than American coal-fired power plants, which average only 32 percent efficiency. Moreover, China is developing not only its coal power but also hydroelectric, wind, and solar technologies. A single Chinese company, Suntech, has become one of the top five producers of photovoltaic cells since it opened its doors in 2001; Suntech's expansion plans alone will exceed all U.S. production by 2010. The Indian company Suzlon has become the fifth-largest global producer of wind turbines in just over five years. The technologies involved originated in the United States, but they are now being produced offshore as a direct result of American policies that disadvantage renewable energy in relation to fossil fuels.

The rest of the world is not waiting for American leadership to address global climate change. And the consequences of our failure to lead will be not only environmental but economic: while we debate, Europe, Japan, India, and China are poised to become the world's suppliers of sustainable-energy technology.

BOSTON REVIEW BOOKS

Boston Review Books are accessible, short books that take ideas seriously. They are animated by hope, committed to equality, and convinced that the imagination eludes political categories. The editors aim to establish a public space in which people can loosen the hold of conventional preconceptions and start to reason together across the lines others are so busily drawing.